ESPIONAGE BLACK BOOK SEVEN

In this series of technical monographs:

Espionage Black Book One: Intelligence Databases Explained

Espionage Black Book Two: Codes and Ciphers Explained

Espionage Black Book Three: Surveillance Explained

Espionage Black Book Four: Open-Source Intelligence Explained

Espionage Black Book Five: Basic Intelligence Explained

Espionage Black Book Six: Spy Tradecraft for Citizens Explained

ESPIONAGE BLACK BOOK SEVEN:

Spy Radio Explained

Henry W. Prunckun

Bibliologica Press

Acknowledgements

Appreciation is extended to those who have provided photographs and diagrams for use in this book. In the case of the American Radio Relay League and Icom Inc., it is pointed out that they retain copyright for the images reproduced herein.

Espionage Black Book Seven:
Spy Radio Explained
by Henry W. Prunckun

ISBN 978-0-6456209-1-7

A catalogue record for this
book is available from the
National Library of Australia

For information on all Bibliologica Press's publications,
visit our Web site at bibliologica.com

Bibliologica Press
P.O. Box 656
Unley, South Australia, 5061
Australia

CONTENTS

— CHAPTER ONE —

SPY RADIO

In February 2022, Vladimir Putin ordered the Russian military to invade Ukraine. History shows that as Putin's armed campaign increased, he tried to eliminate Russia's access to news about the disastrous developments encountered by his invading forces, especially regarding the war crimes they committed.

The United Kingdom's BBC World Service[1] responded by airing two foreign language programs—one in Ukrainian and the other in Russian—to provide accurate news about the situation.[2] Shortwave radio was used for the BBC's radio signals to travel from England to their audiences in Eastern Europe. In this case, the BBC used the shortwave frequency of 5.878 MHz to prove that the bands formerly used by intelligence agencies for long-distance communications were still valuable.

1. BBC World Service is the international broadcasting arm of the British Broadcasting Corporation, which delivers news and content through a range of languages and regional services over radio, television, and digital modes.

2. The broadcasts were reported to have gone to air been between to the hours of 4pm to 6pm and again from 10pm to midnight, Ukraine time. Sian Cain, "BBC Website Blocked in Russia as Shortwave Radio Brought Back to Cover Ukraine War," The Guardian. Found at: https://www.theguardian.com/media/2022/mar/04/bbc-website-blocked-in-russia-as-shortwave-radio-brought-back-to-cover-ukraine-war. Accessed: October 14, 2022.

Although the BBC's 2022 Ukrainian and Russian signals were not sent by spies to spies, the tradecraft that underpinned spy radio practice was the same. It is this tradecraft that is used by espionage operatives.[3] In fact, political broadcasts like those of the BBC have been associated with intelligence agencies' clandestine work since the Second World War.[4]

Yet, there have been other clandestine stations operated by other countries. Take, for instance, the Cold War station *Radio Free Europe* and *Radio Liberty. Radio Free Europe* broadcasted to the Soviet Union's satellite countries, whereas *Radio Liberty's* target audience was the Soviet Union.[5]

Radio Free Europe began as an anti-communist front organization for the CIA, but since 1972 it has had no affiliation with the agency.[6] Formed in 1949 when Allen Dulles was Director of Central Intelligence, *Radio Free Europe* and *Radio Liberty* were financially underwritten by the CIA. Still, although they operated like any

---

3. The term *operative* has drawn criticism from some intelligence practitioners. These clandestine services personnel have cast disapproval on scholars who, for the sake of simplicity, use this generic term rather than list the different the occupational-specific titles. It is not always possible to specify these human intelligence (HUMINT) titles because there are many. These titles include—e.g., case officers, agents, couriers, operations officers, paramilitary operations officers, language officers, specialized skills officers, staff operations officers, and targeting officers.

4. Harry L. Helms, *How to Tune the Secret Shortwave Spectrum* (Blue Ridge Summit, PA: TAB Books, 1981), pp. 7 & 36.

5. John Prados, *Safe for Democracy: The Secret Wars of the CIA* (Chicago: Ivan R. Dee, 2009).

6. Since 1995, the radio stations' headquarters has been in Prague, Czech Republic.

commercial international news agency, the CIA and the U.S. State Department did issue broad anti-communist content policy directives.[7]

How do these clandestine shortwave stations[8] relate to spy radio? What tradecraft links them to other manifestations of covert radio communications? These and other questions that examine the technical basics of radio, receivers and transmitters, antennas, and transmission lines, as well as electromagnetic propagation and countermeasures, will be discussed in the coming chapters. But first, let us look at a few other case studies of how spy radio is used in practice.

Figure 1—A shortwave radio reception report (known as a *QSL card*) from the Japanese doomsday cult, Aum Shinrikyo.

---

7. A. Ross Johnson, *Radio Free Europe and Radio Liberty: The CIA Years and Beyond* (Washington, DC: Woodrow Wilson Center Press and Stanford University Press, 2011).

8. Not to be confused with *pirate radio*, which is "…any sort unlicensed, illegal radio activity intended for reception by the general public." Harry L. Helms, *How to Tune the Secret Shortwave Spectrum*, p. 97.

Intelligence operations that used clandestine radio techniques did not end when the Berlin Wall fell. It is still used by spies in denied areas and a range of radical underground groups.

For example, the Japanese doomsday cult Aum Shinrikyo broadcast delusional dogma in the early-1990s on 17.710 MHz from a radio station under contract in Russia. The daily broadcasts were an hour long and beamed to Japan, but they could be heard world-wide, including from my personal listening post in Australia. Figure 1 shows a copy of the radio reception report I received from the station's management in relation to one broadcast I received.

Since the 9/11 attacks, intelligence tradecraft has expanded to groups that have spun out of extremist religious briefs instead of geopolitical alliances.[9] A case in point is are the terrorists who trained with al-Qaeda and its affiliated groups have joined Russian, Cuban, Chinese, and North Korean agents who were represented as the notorious "opposition" in many of the classic Cold War spy novels.[10]

During the Second World War and the Cold War, deep cover operatives and agents-in-place passed messages to

---

9. See, *Military Studies in the Jihad Against the Tyrants: The Al-Qaeda Training Manual*. This underground training manual was located by the Manchester Metropolitan Police while searching the home of an al-Qaeda member. The manual was translated into English and introduced into evidence at the 2001 criminal trial of al-Qaeda terrorists in New York for the 1998 bombings of the U.S. embassies in Kenya and Tanzania.

10. For instance, the novels by John LeCarre—*The Spy Who Came in from the Cold* (1964), *A Perfect Spy* (1986), *The Russia House* (1989), and the *Secret Pilgrim* (1990), as well as others.

HQ using several methods—dead drops, cut-outs, etc.,[11] as well as secret radio transmissions, were standard operating procedures. These are documented in the scholarly literature on espionage, particularly secret radio transmissions.[12]

In the main, field operatives have communicated using high-frequency (HF) radio transmissions,[13] which are also known as *shortwave* transmissions. Former CIA officer, the late-Miles Copeland,[14] described this practice in some detail in his book *The Real Spy World.*[15] In the days before the Internet and email, HF radio was arguably the most, if not the only, reliable method of world-wide communication *and* the method least likely to be subject to interception—unlike landline and cellular telephone, telex[16] or facsimile that use dedicated and identifiable telecommunications cables that can be "tapped." Copeland explains why:

> Clandestine radio is also safer than is commonly supposed. Even with the most modern detection equipment, frequencies used for espionage transmissions

---

11. Wayman C. Mullins, *Terrorist Organizations in the United States* (Springfield, IL.: Charles C. Thomas, 1988), p. 56.

12. For instance, see Louis Meulstee and Rudolf F. Staritz, *Wireless for the Warrior, Volume 4: Clandestine Radio* (Dorset, England: Wimborne Publishing Ltd, 2004).

13. The high-frequency (HF) range spans the electromagnetic spectrum between 3 MHz and 30 MHz.

14. Miles Copeland (1913–1991) was once a CIA officer who served in the Counter-Intelligence Corps during World War II.

15. Miles Copeland, *The Real Spy World* (London: Weidenfeld and Nicolson, 1974).

16. Although telex is not a common a mode of communication, at the time of writing it was still in limited use.

> are hard to catch—and still harder to identify for what they are because the transmissions sound like ordinary coded messages used legitimately by diplomatic and commercial concerns. Even if they come under active suspicion, by the time direction finders ("DF-ing equipment") are in place to seek the point of origin, the message is over. Naturally, the "DF-ers" listen for the next transmission, but it may be from a different place. Modern security agencies claim that they have sophisticated equipment which will overcome these difficulties, but it simply isn't so. The airwaves are full of illicit transmissions—from China, from various parts of Russia, and from the Third World. When senders are caught, it is usually as the result of tip-offs from suspicious neighbors rather than from successful DF-ing. DF-ing gets the security agents to the right general area, but that is about all.[17]

Intelligence officers who operate overseas under non-official cover (abbreviated NOC, and pronounced "knock"), as well as any agents they recruit, need to communicate with those directing their mission. Operations officers need to receive instructions and send data. Both HQ and operatives also need to ask questions as issues arise.

Before operations officers deploy overseas, they visit their agency's technology directorate to discuss their covert communications (covcom) requirements. While operatives who work from embassies under official cover have the luxury of using secure computer links to transmit their cables, NOCs need other devices. Although a NOC's covcom device is likely to be embedded in a laptop-type computer, it is not always the case. When deployed to

---

17. Miles Copeland, *The Real Spy World*, p. 143.

countries that lack Internet or where the Internet is controlled by the government, other devices are needed.[18]

This is where radio comes to the fore, for example, *numbers stations*. These are one-way transmissions from HQ to field operatives. These radio transmissions are referred to as one-way voice links (OWVL) by intelligence agencies and use shortwave, very-high frequency (VHF), and ultra-high frequencies (UHF) frequencies. It is also understood that beams of infrared light have been used for two-way communications, but due to the nature of light, this technology can only be used for parties that are in sight of each other (i.e., point-to-point).

Numbers stations operate throughout the shortwave bands, as known as the high-frequency bands. These frequencies are used because when the earth's upper atmosphere—the ionosphere—becomes charged by the sun, it reflects the radio's waves back to earth.[19] This phenomenon does not take place if the signals are above 30MHz. In the case of the VHF band and UHF, the signals pass through the ionosphere and into space, so these bands are used for line-of-sight communications (see Table 1 for on these other bands).

The ability to reflect HF radio signals off the ionosphere (known as *skip*) allows government operatives and military units to communicate over very long

---

18. Amaryllis Fox, *Life Undercover: Coming of Age in the CIA* (New York: Alfred A. Knopf, 2019), p. 118.

19. The ionized part of the earth's atmosphere begins around 48 km (30 mi) altitude and continues to about 500 km (310 mi). *The ARRL Handbook* (Newington, CT: American Amateur Radio League, 2001), pp. 21.11–21.17

distances.[20] This is why the HF bands are popular with amateur radio operators; they can make contacts world-wide even with modest equipment.

Numbers stations use the HF frequencies to transmit what appears to be a series of random (or letters), in either voice or using Morse code. It has long been speculated that these stations are operated by government intelligence agencies. The strings of numbers or letters[21] that are sent are hypothesized to be one-way voice links to operations officers and agents overseas, but there has been little in the way of confirmation.

A search of the intelligence studies literature on spy radio shows a lack of detail about this particular clandestine tradecraft. Given that every aspect of spy tradecraft has been written about to one degree or another, it strikes odd that numbers stations have not.

One would expect to see references to numbers stations in exposés, the memoirs of former chiefs-of-station, intelligence directors, military intelligence commanders, and books written by intelligence studies scholars. In his comprehensive treatment of signals intelligence, not even the acclaimed espionage scholar Nigel West,[22] nor James Bamford's in-depth examination of the National Security

---

20. James R. Clapper with Trey Brown, *Facts and Fears: Hard Truths from a Life in Intelligence* (New York: Viking, 2018), p. 25.

21. Harry L. Helms, *How to Tune the Secret Shortwave Spectrum*, p. 49.

22. Nigel West (pseud., Rupert William Simon Allason), *Historical Dictionary of Signals Intelligence* (Lanham, MD: Scarecrow Press, 2012).

Agency, mentions numbers stations.[23] Neither does David Kahn in his definitive work on secret communications, *The Code Breakers*.[24] It would appear that intelligence agencies have shied away from acknowledging their use.[25]

Although, perhaps without realizing it, in 1998, the FBI revealed a group of five Cuban spies operating in Florida (known as the *Wasp's Network*) were arrested, and in doing so, unwittingly made public that these operatives were receiving coded messages from Cuba via shortwave broadcasts. In prosecuting the five Cuban operatives for espionage,[26] the FBI made an oblique admission that OWVLs are used in intelligence work.

There have been other cases where indirect revelations have been made. Take the case of the Cuban spy who penetrated the U.S. Defense Intelligence Agency and the situation of two Russian spies who operated in German.

> Montes, a Cuban spy who worked for the U.S. Defense Intelligence Agency, was convicted of espionage in 2002. She had been passing secrets to Cuba for more than twenty years when she was caught. When agents searched her apartment, they found a small shortwave radio and a piece of paper containing a matrix of numbers and letters that they believe was used as a deciphering pad. The Anschlag case is more recent. In 2011 Andreas and Heidrun Anschlag, [a married Russian couple] who

---

23. James Bamford, *The Puzzle Place: A Report on America's Most Secret Agency* (Boston: Houghton Mifflin, 1982).

24. David Kahn, *The Code Breakers: The Story of Secret Writing* (Toronto: The Macmillan Company, 1967).

25. Louis Kruh, "Shortwave Numbers Stations—Recordings," in *Cryptologia*, Volume 23, Issue 2, April 1999.

26. Allison McLellan, "Decoding Numbers Stations, in *QST*, November 2019, p. 73

> had been living in West Germany for more than twenty years under assumed identities, were arrested. According to media reports, when a Special Forces commando stormed the Anschlag's house he caught Heidrun in the act of receiving a coded message on a shortwave radio.[27]

Nevertheless, there are a few authoritative (though not official) references in the literature that acknowledge the use and purpose of numbers stations. The closest to an official acknowledgement is from Robert Wallace and H. Keith Melton's book, where the two espionage scholars (Wallace is also a former director of CIA's Office of Technical Services) stated that spies received "...a string of random numbers[28] over a standard shortwave radio, and decrypted those numbers with OTPs [one-time pads] like those used by the French and Polish underground in Occupied Europe."[29] They went on to describe this method in detail.

> The one-way voice link described a covert communication system that transmitted messages to an agent's unmodified shortwave radio using the high-frequency shortwave bands between 3 and 30 MHz at a

---

27. Christopher Friesen, "Spy 'Numbers Stations' Still Enthrall," in *Radio World*, Volume 38, Issue 2, January 15, 2014, p. 14.

28. Although they refer to the message as "random numbers," these were not random numbers. What was meant, is that the numbers appeared to be random but were in fact an encrypted message expressed as what would appear to the novice listener as random numbers. See, Henry W. Prunckun, *Espionage Black Book Two: Codes and Cyphers Explained* (South Australia: Bibliologica Press, 2021).

29. Robert Wallace and H. Keith Melton with Henry Robert Schlesinger, *Spycraft: Inside the CIA's Top Secret Spy Lab* (New York: E.P. Dutton, 2008), p. 85.

> predetermined time, date, and frequency contained in their communications plan. The transmissions were contained in a series of repeated random number sequences and could only be deciphered using the agent's one-time pad. If proper tradecraft was practiced and instructions were precisely followed, an OWVL transmission was considered unbreakable. The agent was able to use OWVL only to receive communications, but it had many advantages over secret writing or agent meetings. OWVL required no spy gear except a one-time pad, was generally reliable and repeatable, and precluded surveillance. As long as the agent's cover could justify possessing a shortwave radio and he was not under technical surveillance, high-frequency OWVL was a secure and preferred system for the CIA during the Cold War.[30]

There is also a detailed discussion of how Soviet "illegals"[31] used OWVL in the late-Oleg Penkovsky's book, *The Penkovsky Papers*.[32] In brief, Penkovsky was a Soviet military intelligence colonel during the late-1950s and early-1960s who worked as a double agent for Western intelligence. Although his book reveals much about the intelligence operations and tradecraft of Soviet intelligence, it was later discovered that the book was written by the CIA using "...witting Agency assets who

---

30. Wallace and Melton with Schlesinger, *Spycraft*, p. 498.

31. " In KGB parlance, an illegal was a case officer living in a target country under false identity and no apparent connection to the USSR." Ted Shackley, with Richard A. Finney, *Spymaster: My Life in the CIA* (Dulles, VA: Potomac Books, 2005), p. 292.

32. Oleg Penkovsky, *The Penkovsky Papers* (London: Collins, 1965), pp. 81, 311–312, 320.

drew on actual case material,"[33] thus providing another indirect confirmation OWVLs.

The only official document related to one-directional communications is understood to be a KGB radio manual that was released by the Latvian National Archive. The manual was obtained by the Latvians, along with other classified intelligence documents, after Latvia declared independence from the Soviet Union in 1990. The text explains how KGB counterintelligence officers monitored the clandestine OWVL broadcasts of the CIA and the of what was then, West Germany's Federal Intelligence Service (BND).[34]

Even though there have been some scholarly attempts to improve one-way voice links,[35] numbers stations

---

33. Frank Church, *Book I: Foreign and Military Intelligence: X. The Domestic Impact of Foreign Clandestine Operations: The CIA and Academic Institutions, The Media and Religious Institutions, Appendix B* (Washington, DC: U.S. Government Printing Office, April 23,1976), p. 194.

34. Māris Goldmanis, "Numbers Stations Over Soviet Latvia: KGB Counterintelligence Monitoring of Western Intelligence Radio Transmissions." Available at: https://www.numbers-stations.com/articles/kgb-numbers-stations-archive/ Accessed March 29, 2022.

35. For example, in his paper, *A Proposed System for Covert Communication to Distant and Broad Geographical Areas* (2014), Joshua Davis describes a covert communication system "...that modulates Morse code characteristics and that delivers its message economically and to geographically remote areas using radio and EchoLink." He wrote that his "...system allows a covert message to be sent to a receiving individual by hiding it in an existing carrier Morse code message. The carrier need not be sent directly to the receiving person, though the receiver must have access to the signal. Illustratively, we propose that our system may be used as an alternative means of implementing numbers stations." p. 1.

remain an efficient and effective way of communicating with operatives in "denied areas" (i.e., locations where it is challenging to use more open forms of communications.

Although numbers stations vary in their operating procedures, there is a typical format. Transmissions usually begin on the hour, half-hour, or quarter-hour.[36] These transmissions begin with preliminary information that identifies the station and the recipient, just as amateur radio stations do.[37] For instance, a numbers station prelude might be: "Attention, Attention, Attention, 176 29."

In this example, the transmitting station may repeat this prelude several times, giving the receiving station time to tune their equipment and get pencil-and-paper ready. There is some debate about what the first three numbers indicate. Some scholars assert it refers to the operative to whom the message is directed; others posit that it is a code key for the message.[38] The last two digits advise that there will be twenty-nine number groups in the message. This information acts as a check for the operative to ensure that a group or groups weren't missed.

The language may be in English, Spanish, Russian, German, or another tongue. The message may be sent as instructions or as a training message. There is some speculation that phantom messages are sent amongst

---

36. Wallace and Melton with Schlesinger, *Spycraft*, p. 439.

37. Amateur radio is both a hobby and a community service. Amateur radio operators must have a government-issued license. Licenses are issued after a candidate passes written exams covering legal regulations and electronic theory. They must also sit a practical exam on correct operating procedures.

38. Harry L. Helms, *How to Tune the Secret Shortwave Spectrum*, p. 41.

genuine transmissions to confuse and mislead listening opposition intelligence services.

The format of the messages can be in three-, four- or five-digit blocks; they can consist of numbers or letters, and the encrypted message may be repeated twice during the transmission. Repeating the message allows the receiver another check that all was received.

Overall, it is a slow process that requires concentration and attention to detail. So, it is not surprising that as technology advanced, starting in the 1970s, the CIA's Office of Technical Services and the Office of Communications began redeveloping the agency's OWVL system more efficiently. The time-tested shortwave receiver was substituted with a:

> ...dedicated IOWL [interim one-way link] receiver. The self-contained miniature piece of spy gear was a black box about the size of a pack of cigarettes and half as deep, including the internal battery. Its size made concealment relatively easy, and it could be plugged into a standard speaker or operated with headphones. The primary benefit to the agent was the speed of receiving a message; the numbers were transmitted at higher speeds and then stored internally in the receiver to be recalled later. Decreasing the time an agent had to spend performing the covert activity of listening to and transcribing shortwave transmissions improved security and efficiency; messages that previously had required the agent to listen and copy for an hour could be received in ten minutes. IOWL required the agent to possess and hide another piece of spy gear, but because it was technically equal to OWVL and offered advances in

reception speed and an improvement in weak-signal reception, the system was widely deployed.[39]

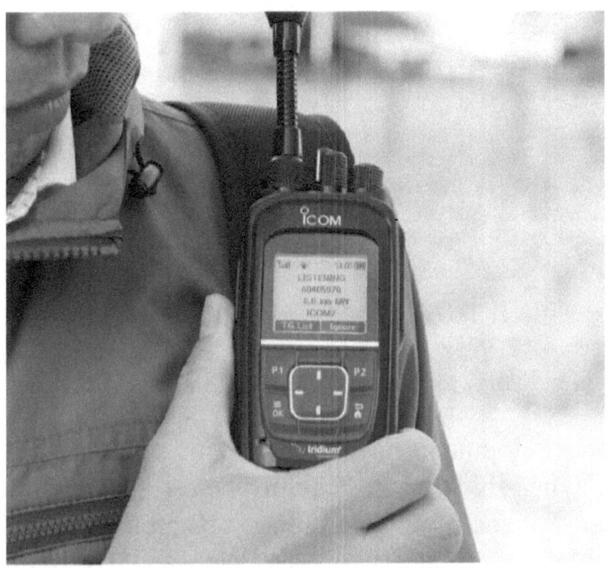

Figure 2—An example of a commercial satellite transceiver using the Super-High Frequency (SHF) band. Courtesy of Icom Inc.

At the time of this writing, the advances in wireless technology have grown far beyond the CIA's dedicated OWVL receiver. Arguably, microchips and software are what make modern radio receivers (i.e., known as *software-defined radio* and abbreviated SDR). Software-defined radios have replaced shortwave receivers using electronic components mounted on circuit boards (such as those shown in Figure 7) because of their versatility and the many features they offer.[40]

---

39. Wallace and Melton with Schlesinger, *Spycraft*, p. 499.

40. In terms of versatility, a SDR receiver can be setup in a target city and connected to the Internet so that the signals can be monitored/recorded a world away. SDRs also offer the radio

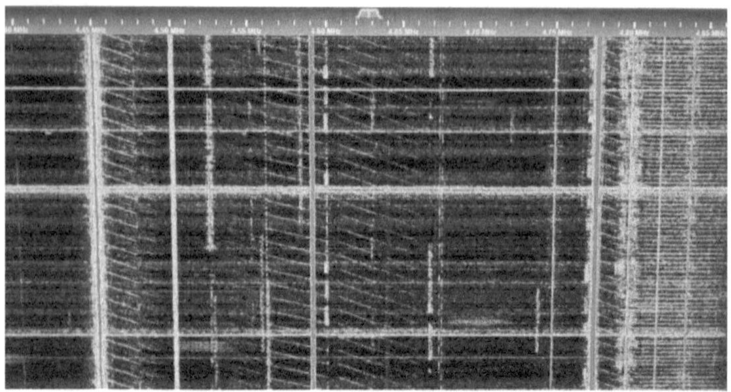

Figure 3—A radio operator's view of the radio spectrum between 4.40 MHz and 4.85 MHz as presented on an SDR's computer display as a spectrogram. Author's collection.

Other developments have been in using satellites as a way of relaying signals across hemispheres rather than using the ionosphere to "bounce" HF signals around the globe.[41] Satellites also make two-way voice and data contacts possible using the VHF, UHF, and SHF bands (see Table 1 for a description of these bands). This is because these frequencies pierce the ionized layers of the upper atmosphere.

Essentially, operatives send their signal skywards, thereby avoiding radiating their radio signal to surrounding areas. Limiting their signal's area of coverage reduces the possibility of being intercepted by the opposition and makes it difficult for them to triangulate the operative's position using direction-finding techniques. Moreover, because they are no physical objects in the path of a signal as it "beams" its way aloft,

---

operators the ability to view a wide portion of the radio spectrum via a display known as a *spectrogram*.

41. Also known as *skip*.

lower-powered handheld covcom devices can be used (see Figure 2 for an example of a commercial version).

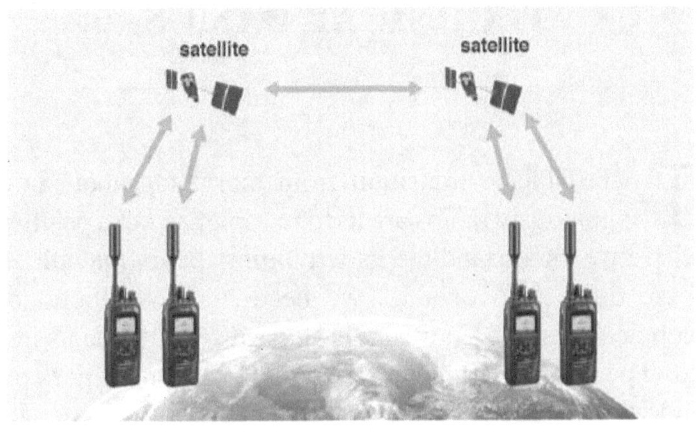

Figure 4—A demonstration of how satellite transceivers can be used for world-wide communications without the need of a shortwave transceiver. Courtesy of Icom Inc.

There are many shortwave radio listeners who have made it a hobby to monitor numbers stations and record their coded messages. At the time of this writing, there were several websites that posted information about spy signals, yet the scholarly literature was barren on tradecraft details. Perhaps this is due to scholars' lack of understanding of the technical aspects of radio?[42] This is understandable because while theory is straight-forward, the actual covcom systems are scientifically advanced. To understand more, let's look at the technical basics of radio.

---

42. By way of example, in *Spies and Lies: How China's Greatest Covert Operations Fooled the World* (London: Hardie Grant, 2022), political espionage scholar Alex Joske discussed how clandestine operatives in China communicated with agent-handlers with nothing more than, "The CIA has its own, more secure systems for sending messages back to headquarters." p. 116

— CHAPTER TWO —

TECHNICAL BASICS

To be able to transition from marveling about how ingeniously radios are used in espionage to a position where we understand the underpinning theory that allows these devices to operate, we need to know about the technical basics. Knowing these basics will place us in a good position to appreciate what radios can make possible and what they are not likely to do.

Suppose we apply the analogy of the technical basics of a car. In that case, we can appreciate what particular cars are capable of and what they are not—we can separate fact from fiction and, in some cases, fantasy.

In the following chapters, we will discuss the basics of spy radio—receivers and transmitters, antennas and transmission lines, propagation, radio interference, and electrical safety. But first, we will discuss the foundation on which all electronic equipment is based—electricity.

Electricity is a physical phenomenon. It exists in two forms—static and current.[43] We are familiar with static electricity because we often receive a "zap" when brushing our hair, rubbing our feet on a carpet, or brushing against certain types of clothing. We also see displays of

---

43. W. Thomas Griffith and Juliet W. Brosing, *The Physics of Everyday Phenomena, Seventh Edith* (New York: McGraw-Hill, 2012).

static electricity in the form of lightning when clouds discharge a build-up of charged particles.

We are familiar with current electricity because we use it to power everything in our life—lights, kitchen appliances, tools, cars, and all our smart devices.

Yet, we cannot see, smell, hear, or taste electricity; so, what is it? Simply, electricity is the orderly movement of *electrons* in a conductor. Physicists call this phenomenon an *energized circuit*. It is this energy (i.e., the moving electrons) that provides the power to allow a *load*[44] to do something—light a light, power a toaster or drill, charge a car battery, or allow our smart devices to work.

Figure 5—Static electricity in the form of lightning. Courtesy of the U.S. Air Force.

In the main, we used electric currents in our everyday lives, and this comes in two forms—direct current and

44. A *load* refers to the power consumed by a device or circuit while performing its function. Rudolf F. Graf, *Modern Dictionary of Electronics* (Slough Bucks England: W. Foulsham, 1967), p. 201.

alternating current. Direct current (DC) is current that flows in one directions, from negative to positive. Alternating current (AC) flow through a conductor in both directions—first one way and then in the opposite direction, hence the name *alternating*. DC currents are used in battery-powered devices, whereas AC is used to power devices that are connected to the power grid. That is, the electricity generated by solar farms, wind turbines, and nuclear and gas/coal-fired power plants.

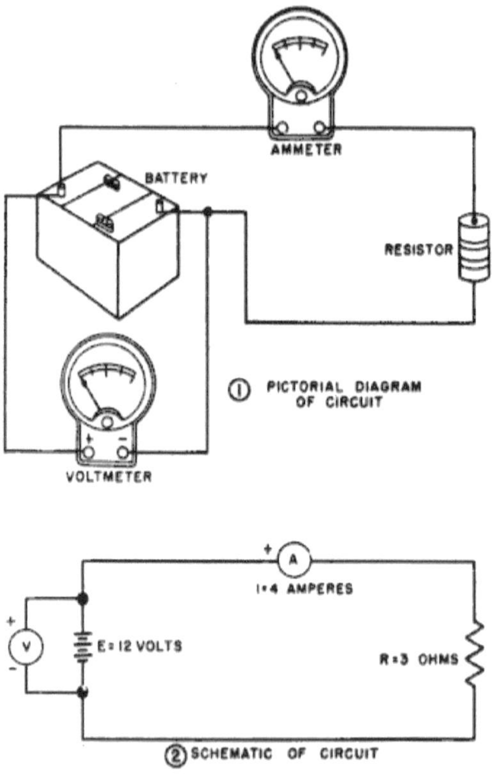

Figure 6—An example of a pictorial diagram and a schematic diagram of a basic electrical circuit. Courtesy of the U.S. Air Force.

Current electricity needs a means to move electrons from the energy source (i.e., a battery or generator) to the load. This is accomplished by using conductors. All metals are conductors of electricity, though copper and aluminum are the most common metals in electric circuits.

To ensure that the energy (i.e., moving electrons) passes through conducting wires—whether it is a few inches, as in an electric circuit board, or transcontinental, in the case of AC transmission lines, insulators are used to isolate the conductor from other conductive paths.

This is important because if the energy in the conductor is diverted from its journey from source to load, a *short-circuit* will occur. Depending on the amount of electrical energy that is passing through the conductor at the time of a "short," it could damage the device, overheat the wires, and start a fire, or if it touches a person, cause injury or death.[45]

Insulators can be made of a variety of materials. For flexible wires, it can be PVC, plastic, or rubber compounds. In fixed positions, it can be glass, ceramics, or hard plastics. Because of their physical properties, these, as well as others (e.g., dry wood), do not allow electrons to pass through them, thus allowing the moving electrons to stay within the circuit.

Although electricity is not tangible, we can measure moving electrons using two metrics—voltage (V) and amperage (A). If we use the analogue of water flowing through a pipe, voltage represents the pressure exerted

---

45. Peter E. Sutherland, *Principles of Electrical Safety* (Hoboken, NJ: John Wiley & Sons, 2015).

(force), and amperage represents the amount of the flow (volume).[46]

As current flows through a wire, it meets resistance (R).[47] These wires, because of their narrow cross section, offer more resistance and hence restrict the flow of electrons. Larger diameter wires allow current to flow with less restrictions and, therefore, can handle larger volumes of current.

The relationship between voltage, amperage and resistance is expressed in Ohm's Law. Ohm's Law is a mathematical equation that states that voltage divided by resistance equals amperage. When expressed in these terms, the symbols for voltage change to the capital letters E and I, respectively.

$$E / R = I$$

Ohm's Law can be rearranged to solve for voltage (E) and resistance (R), like this:

$$I \times R = E \quad \text{and} \quad E / I = R$$

A quick example of how Ohm's Law is used is where we, say, have a 1.5V battery and a 1-ohm resistor is placed across the terminals, a 1.5 ampere current will flow.

The reason we need to understand these relationships is that they allow radio engineers to design electronic circuits that function as radios. Of course, there are other mathematical formulae, as well as additional physical

---

46. One ampere (an "amp") is one coulomb (C) of electrons per second. One coulomb is equal to approximately 6.241 x 10^{18} electrons.

47. Resistance is measured in *ohms*.

laws and theories that must be mastered to be able to create wireless devices.[48] Nevertheless, just this basic understanding of circuits will give us the understanding we need to appreciate what is possible for radios to do and what they are not likely to be able to perform.

Other than to briefly list a few of the most common components used in electronics in Appendix A, we will leave that discussion and move on to explaining circuit diagrams. Our aim is not to become radio engineers, but to gain basic understanding of how radios work so we can apply that knowledge to provide insights into the operational parameters of spy radio. In this regard, *frequency* is the next concept we will discuss.

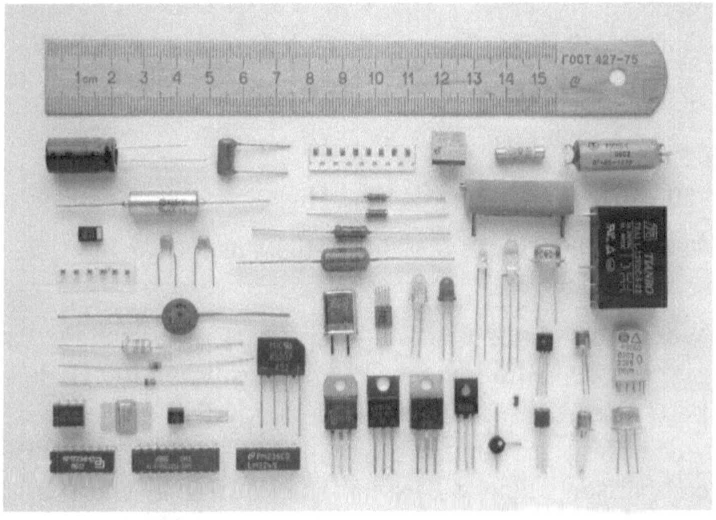

Figure 7—A selection of electronic components—the types that are found on the circuit boards of radios.

---

48. Nathan Blaunstein, Christos Christodoulou, and Mikhail Sergeev, *Introduction to Radio Engineering* (Boca Raton, FL: CRC Press, 2016).

Frequency describes the number of times something happens, as in the number of times an alternating current changes direction, the number of oscillations associated with a radio signal, or the number of "waves" in a particular sound. Frequency (*f*) is measured in Hertz (Hz). For instance, in the United States, the frequency of AC current is 60 times per second or 60 Hz. The human ear can hear sounds as low as 50 Hz and as high as 1,500 Hz (or, 1.5 kHz).

Table 1—Common radio bands.

Frequencies	Band	Abbreviation
3 kHz–30 kHz	Very Low Freq.	VLF
30 kHz–300 kHz	Low Freq.	LF
300 kHZ–3 MHz	Medium Freq.	MF
3 MHz–30 MHz	High Freq.	HF
30 MHz–300 MHz	Very High Freq.	VHF
300 MHz–3 GHz	Ultra-High Freq.	UHF
3 GHz–30 GHz	Super High Freq.	SHF

When transmitting intelligible voice information via radio,[49] the audio frequencies are narrower—300 Hz to 3 kHz. Yet, the electromagnetic energy that carries these voice messages can, in theory, be of greater range. By comparison, radio waves can span 30 Hz to 300 (GHz) (gigahertz),[50] though the most useful frequencies sit between 300 kHz and 3,000 MHz. The AM broadcast band, for instance, uses the frequencies between 540 kHz

---

49. Referred to as *radiotelephony*.

50. Douglas B. Miron, *Small Antenna Design* (Boston: Newnes, 2006).

to 1,600 kHz (1.6 MHz), and is referred to as *medium frequency* band. Whereas FM radio stations use frequencies between 88 MHz and 108 MHz, which lie in the *very-high frequency* band (3 MHz to 30 MHz).

Sometimes radio engineers refer to a group of frequencies as a *band* that is defined by its wavelength. An FM radio station that operates on 99.9 MHz is in the 3-meter band. That is to say, the wavelength of 99.9 MHz is approximately 3 meters in length.

An AM radio station operating on 729 kHz operates in the 411-meter band. The relationship between wavelength and frequency is expressed in the formula: velocity divided by wave frequency equals wavelength.

$$v / f = \lambda$$

Velocity (v) is expressed as the speed of light because that is how fast radio waves travel (i.e., 300 million meters per second), frequency (*f*) is expressed in megahertz, and wavelength (λ) is expressed in meters. So, taking our earlier example of the FM radio station on 99.9 MHz, we were able to calculate its wavelength thus:

$$300 / 99.9 = 3$$

Why is it important to know this? Because the wavelength tells us how long an antenna needs to be to radiate the radio signal efficiently. Likewise, this is the length a receiver's antenna needs to be to pick up the incoming signal.[51]

---

51. To some degree all antenna designs are a compromise. There is the inevitably a trade-off between RF gain, directivity, mono- or multi-band, physical size, weight, installation requirements (or

When we see a hand-launched battery-powered drone flying overhead, we can judge that the operator is close-by and the vision being captured, if relayed by radio, is likewise only able to be transmitted a short distance. We know this because the size of the antenna the drone has is very small—perhaps a few centimeters long.

Reflecting on our discussion of radio bands, we know that a frequency of a few centimeters would lie in the UHF band and UHF frequencies travel in a straight line. The ionosphere does not reflect radio waves at these frequencies, so the ground control station must within sight of the drone.

We can also deduce that given the small size of the battery, it must be drawing a sizeable amount of current (A) to power the electric motors, the flight controller, and onboard camera. Therefore, the flight time will not be long.

Another example is the ubiquitous handheld transceivers we see emergency and first responders, soldiers, and other combatants using in the field. How is a spy able to determine who there are talking to? The antenna is the clue. A very short antenna suggests UHF band frequencies, and a bit long antenna suggests VHF band frequencies.

In the case of UHF, we would conclude a short distance or perhaps five kilometers. With VHF frequencies, a little longer distances are achievable, maybe ten to fifteen kilometers. If, however, the antenna is very long, it would indicate lower frequencies are being used—those in the

---

portability), cost, and, in undercover field operations, the need for stealth. Marko Suojanen, *Military Communications in the Future Battlefield* (Boston: Artech House, 2018), pp. 95–98.

HF band. In this case, longer distances are likely, perhaps one hundred, or more.[52]

Figure 8—USDA employee using a UHF handheld transceiver. Photograph by U.S. Department of Agriculture, courtesy of National Archives and Records Administration.

---

52. The potential distance a radio signal can travel is to some degree related to the power generated in the device's final output stage. Power is measured in watts and the rule-of-thumb, more power radiated at the antenna increases the likelihood that the signal will be *intelligible* at the receiving end. Note, that more power won't necessarily cause the signal to travel further, but higher power will help the receiving radio's circuitry to distinguish the desired signal from atmospheric noise and/or adjacent signals. From instance, I have made contracts with stations in Antarctica, New Zealand, and Wake Island from Australia, as well as many interstate contacts, using only five-watts of power. Low power stations are referred to as QRP operations. See, George Dobbs, *QRP Basics* (London: Radio Society of Great Britain, 2012).

— CHAPTER THREE —

RECEIVERS AND TRANSMITTERS

Radio devices can be categorized as either *receivers* or *transmitters*. There are, of course, many dozens of peripherical devices that complement radios and how they work, but for our purposes, we will limit our discussion to these two categories. Still, we will also include a hybrid device that comprises both a receiver and transmitter, known as a *transceiver*.

Receivers are intended to collect electromagnetic waves from the space that surrounds us. We do this by using a conductive material cut (i.e., "tuned") to the wavelength of the signals we want to receive. This is what is called an *antenna*. Antennas can also be tuned broader to cover a frequency band, say, the FM broadcast band, rather than the frequency of a signal FM radio station.

Nevertheless, even an antenna tuned to a frequency band can receive signals outside that band but at a reduced signal strength. This is the case with the AM broadcast radio in your car. This band covers frequencies between 530 kHz to 1,700 kHz, so a half-wave antenna[53] tuned to the middle of this band (e.g., 1,115 kHz) would be about 134 meters long (about 434 feet).

Because no motor vehicle could support an antenna of this length, car manufacturers install "compromise

---

53. We will discuss half-wave antennas and other antennae designs in Chapter Four, "Antennas and Transmission Lines."

antennas."[54] These shorter antennas will still receive AM signals but with a reduced signal strength. While listening to an AM radio station, it will appear loud because commercial radio stations use high-power transmitters, thus compensating for the lack of antenna reception. Radio manufacturers offset the anticipated degradation in signal strength by building into the receiver's design amplifying circuits and other signal improving circuits. We will not go into detail other than mentioning that these circuits exist.

An antenna is an electrical conductor—usually made of aluminium tubing or copper wire. Tiny radio signals from a transmitter make the antenna oscillate, which induces electrical energy into the antenna. This minute amount of energy passes along a wire or cable, known as a *transmission line*,[55] to the receiver. The radio receiver takes this radiofrequency energy, tunes out unwanted frequencies, converts the electrical power of the radio signal into audio energy, and plays these sounds (e.g., music and voices) through a speaker or earphones.

If anyone has looked inside a radio receiver, the configuration of the components and circuitry can be bewildering. This does not stop us from understanding the process of converting the electrical signals picked up by an antenna to audio. To explain this, Figure 9 shows the process using a block diagram. In a radio receiver, each block will be constructed of many circuits to accomplish the task, but here we are not concerned about that; we want to understand the basics.

---

54. Joe Carr, *Receiving Antenna Handbook* (Solana Beach, CA: High Text Publications, 1993), p. 19.

55. Also called a *feedline*.

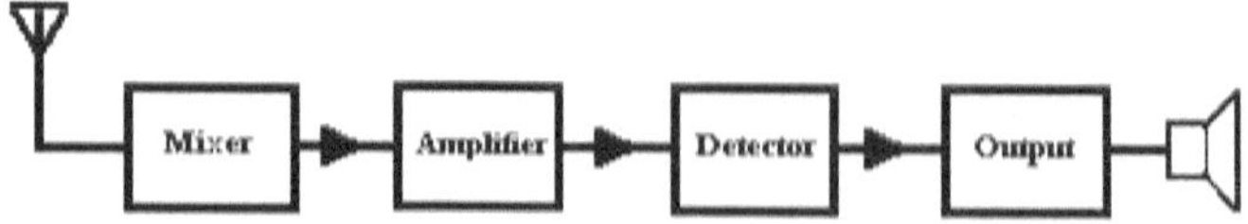

Figure 9—Block diagram of a radio receiver.

Figure 9 simplifies what happens inside a receiver.[56] The antenna is shown on the left. The gathered are conveyed to a "mixer" that combines the incoming signals with a local oscillator signal to produce an "intermediate frequency" that can be amplified in the next stage (i.e., the "IF amplifier"). The signal is then passed onto a detector circuit that converts the RF energy into audio frequency (AF) energy. The next stage—the "output"—is usually an audio amplifier that increases the power to a level where we can hear the sounds through a loudspeaker (shown in the diagram) or a set of headphones.

Where the transmitted signal is digital—as in the case of video or data—the circuitry in the detector stage would be designed to convert the RF energy back into the original digital form rather than audio. The digital data is fed into a device that uses the data—e.g., a video display or computer-readable information—instead of a speaker or headphones.

An example of a digital transmission is a signals sent to a drone directing it to fly in a specific pattern, perform particular tasks, or focus its onboard cameras to pan, tilt, or zoom to capture an image on the ground.

---

56. R.H. Warring, *Making Transistor Radios: A Beginner's Guide* (London: Lutterworth Press, 1970), and Lyle Russell Williams, *The New Radio Receiver Building Handbook* (N.P.: The Alternative Electronics Press, 2006).

At this point, we could delve deeper into the technical aspects of radio receivers. Still, as the purpose is to explain spy radio, we'll limit our discussion to just three related concepts—sensitivity, selectivity, and stability—because these issues also impact transmitters.[57]

Sensitivity describes the receiver's ability to receive weak signals. This is done via the device's radio frequency amplifier and related circuitry. Shortwave (HF) receivers can receive signals as weak as 0.1 uV (microvolt).

Selectivity is important because the radio can select a specific transmission that may be close to other radio transmissions. Figure 10 shows this concept diagrammatically.

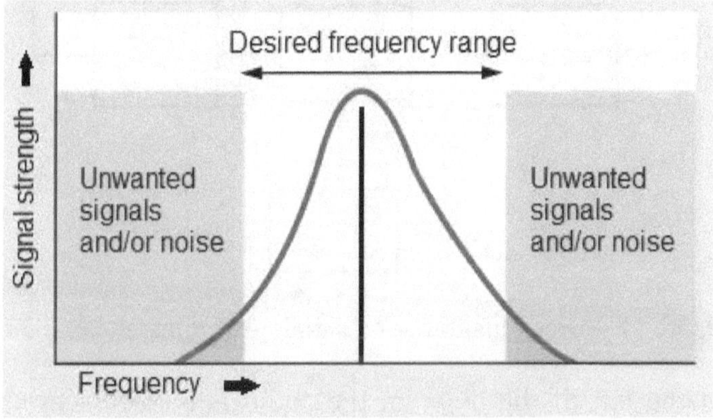

Figure 10—Frequency selectivity by "notching" adjacent unwanted signals and/or atmospheric noise.

Stability relates to the radio's ability to remain on the desired frequency without drifting. Frequency drift can

---

57. Thomas A. Adamson, *Electronic Communications, Second Edition* (Albany, NY: Delmar Publishers, 1992), pp. 95–103.

occur because of the electronic components in the oscillator circuity heat (due to the flow of electrons).[58] Radio engineers can enhance stability by introducing a "crystal" that controls the frequency to very high tolerances; in some cases, less than ±0.5 parts per million (i.e., less than 0.00005 per cent).

Transmitters work in reverse. Transmitters induce an electromagnetic current into an antenna, which oscillates at the desired frequency. These oscillations then travel through space, where we can receive them with a receiver. A block diagram of a simple AM transmitter is shown in Figure 11.

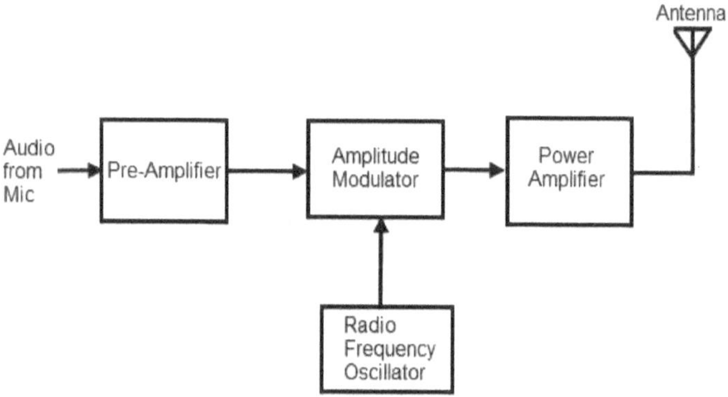

Figure 11—Block diagram of a simple AM transmitter.

At the left of the diagram the audio is collected by the microphone and is sent to a pre-amplifier stage that boosts the modulating signal. At the same time a parallel radio frequency circuit generates a carrier signal by means of an oscillator (often using a high stability crystal).

---

58. Ron Bertrand and Phil Wait, *Your Entry into Amateur Radio, Third Edition* (Bayswater, Victoria: Wireless Institute of Australia, 2016), p. 17.

The audio and the carrier signals[59] are combined in an AM modulator stage and the output is increased in strength in the power amplifier circuit, after which the AM wave is sent to the antenna to be radiated into free space (i.e., transmitted).

---

59. A *carrier* is a wave upon which audio frequency modulation (e.g., voice or music) is superimposed. Eamon Skelton, *Building a Transceiver* (London: Radio Society of Great Britain, 2015).

— CHAPTER FOUR —

ANTENNAS AND TRANSMISSION LINES

Antennas are analogous to tuning folks. They vibrate or oscillate by a current induced in them at the frequency that they are cut (i.e., constructed). This is called the antenna's *resonant* frequency. The function of an antenna[60] is to receive electromagnetic energy sent from a transmitter or transmit this energy to a receiver. This electromagnetic energy is transported from the antenna to the radio receiver, or from the transmitter to the antenna, through a feedline.

If this is the case, why are there so many different types of antennas that we see? Yes, there are many variations of antennas, some that look very odd to the simple, appearing as vertical spears or poles.

The most straight-forward answer is that they are all variations of one design—the *dipole*. Of course, radio engineers could argue differently, but this is enough to explain the theory as it applies to spy radio.

As for transmission lines, although the concept is simple, the theory is more involved, so we will keep the explanation uncomplicated.

An antenna can be used to receive and transmit. Generally speaking, the higher the antenna and the further

---

60. Antennas are sometimes called *aerials* because there are mounted high and clear of surrounding objects.

it is away from human-made constructions (as well as trees, rock formations, etc.), the more efficient the antenna will be. Also, the further away from human-induced electromagnetic interference,[61] the better the signal reception will be for signal reception.

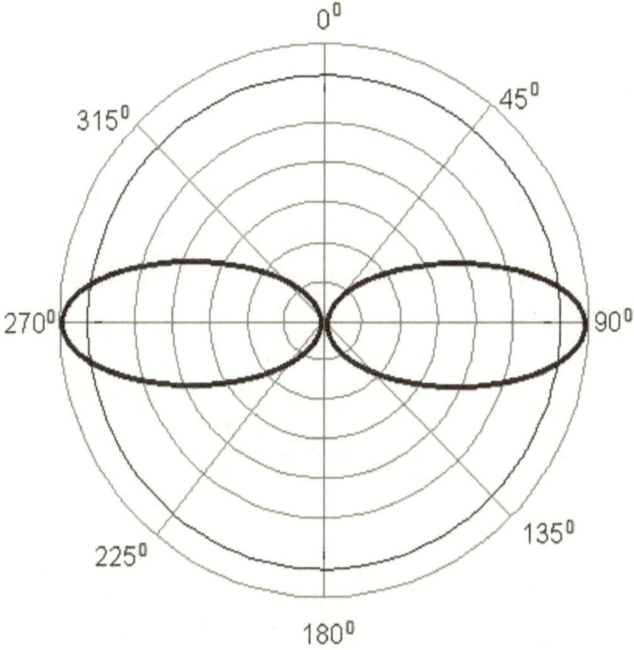

Figure 12—Representation of the horizontal radiation pattern of a half-wave dipole antenna.

Although an antenna has a resonant frequency (or is resonant for a frequency band), it will collect energy for other frequencies. And, if these other unwanted signals

---

61. Electromagnetic interference (EMI) can come from LED lights, plasma televisions, washing machines, battery chargers, ariel power lines, consumer electronics, automobiles, and many other sources. See, Joseph J. Carr, *The Technician's EMI Handbook: Clues and Solutions* (Boston: Newnes, 2000).

have higher energy levels than the chosen signal, they will "drown" the desired signal. These unwanted signals form part of what is known as *background noise*.

The dipole is a half-wave conductor with a feedline connected at its center. It can be mounted either horizontally or vertically. A horizontally mounted configuration is shown in Figure 12. The dipole is a "broadside" antenna because it radiates its energy perpendicular to the conductor. In the case of Figure 12, the radiating element (e.g., a length of wire) is orientated from 0° to 180°. As a result, the RF energy is emitted from the conductor's "sides (i.e., from 270° to 90°), but little energy from the antenna's ends. This feature makes the antenna bi-directional.

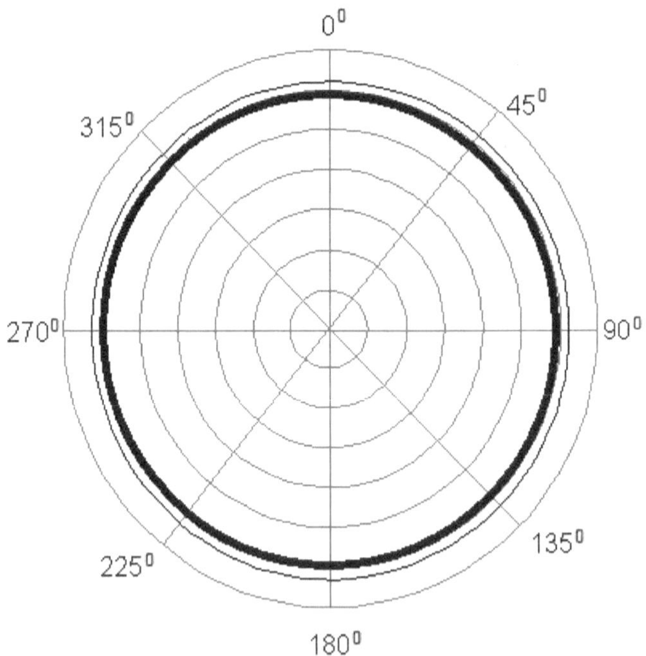

Figure 13—Representation of the horizontal omni-directional radiation pattern of a vertical ground-plane antenna.

The other basic antenna design is the ground-plane antenna. Its design is mounted vertically and has an omni-directional radiation pattern as seen in Figure 13. Its characteristics are a quarter-wavelength long *radiator* and a set of ground plane radials.

It suffices to say that all other variations of antennas, including satellite microwave "dishes," are variations of these two designs. The variations improve directivity (i.e., increase signal strength for both receiving and transmitting) and reject unwanted signals (decreasing spurious signals[62]).

Before discussing feedlines, we will look at another important antenna design and application aspect: *polarization.*

Radio frequency energy combines electrical and magnetic fields, hence the term *electromagnetic.* Figure 14 shows the relationship between the two forces. The figure shows that the two fields are at ninety degrees to each other. In practice, this means that if an antenna is mounted horizontally, it will limit the energy received from a signal emitted from a vertically mounted antenna, in which electrical and magnetic fields will be ninety degrees out of phase.

In this regard, it is important that antennas that use line of sight are polarized the same way. For instance, VHF and UHF frequencies use line of sight. Polarization is immaterial with the bands that reflect their signals off the

---

62. Spurious signals originate from outside the band being used. They can be caused by harmonics generated by radio users operating on other bands, or electromagnetic interference (EMI) from human-made sources (e.g., alternating current [AC] wiring or consumer electronics).

ionosphere because the reflecting action ("skip") stirs the emitted signal's phasing, so either antenna orientation is acceptable.

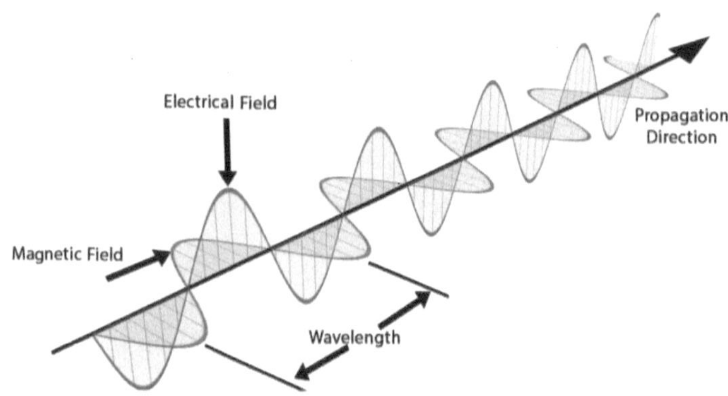

Figure 14—Representation of an electromagnetic wave. Courtesy of Wikimedia Foundation/Dechammakl.

Feedlines, transmission lines or simply *feeders*, are either single wires or a group of wires called *cables*. These lines connect a transceiver to the antenna, and when installed correctly, they do not radiate any RF energy.[63] This is important because if the feedline radiates, some of the signal will be lost (i.e., dissipated through the feedline) on its way to the antenna or from the antenna to the radio. It is essential to have a correctly installed feedline because if it emits radiation, this RF radiation can cause harm to the radio operator.[64]

---

63. M. Walter Maxwell, *Reflections: Transmission Lines and Antennas* (Newington, CT: ARRL, 1990).

64. American Radio Relay League, *The ARRL Antenna Handbook, 21st Edition* (Newington, CT: ARRL, 2009), pp. 1-17 to 1-24.

— CHAPTER FIVE —

PROPAGATION

Propagation refers to how radiofrequency energy travels through space, from a transmitting antenna to a receiving antenna. RF energy is like light energy; both are electromagnetic but differ in frequency. Like light waves, radiofrequency waves travel in a straight line. And analogous to light energy, radio energy can be reflected, refracted, and diffracted.

Line-of-sight (LoS), reflection, refraction, and diffraction are important issues for spy radio. Knowing how radio waves propagate allows technicians to correctly choose the right espionage radio equipment for a particular mission. Mission factors that are considered by technicians are based on "…how frequently the covcom would be used, the size and aggressiveness of the local counterintelligence service, level of surveillance directed against the [operative], and the number and types of covcom systems already operating in the area."[65]

To start our discussion of propagation, we will examine some of the fundamental theories that buttress the way radio is used in practice. The first that we will look at is *field strength.* This is a term that refers to the power of the received signal from the transmitted source. As radio waves travel from their transmitting antenna, their power

65. Wallace and Melton with Schlesinger, *Spycraft*, p. 421.

diminishes as they fill "free space" at the inverse square of the distance.

In lay terms, if the distance between the two antennae doubles, the received signal strength will decrease by a quarter. If the distance triples, the signal strength will be reduced by one-ninth, and if the distance increases to ten times, the signal strength will diminish to one-hundredth.[66]

From the point of view of spy radio, this theory has two applications: the first is that for a message to be received, there needs to be sufficient power in the signal for the receiver to detect it.

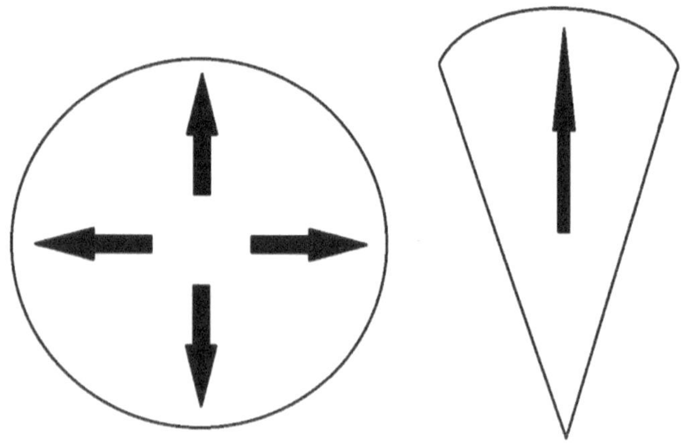

Figure 15—On the left is an models an antenna with an omni-directional pattern and on the right shows the main radiation "lobe" of a directional antenna.

---

66. Ron Bertrand and Phil Wait, *Your Entry into Amateur Radio: The Foundation Licence Manual, Third Edition*, p. 33.

The second point is that much of the RF energy is lost if the signal is being radiated in all directions (e.g., via an omni-directional antenna). This last point could also mean that an opposition's listening post could collect these signals for analysis. Therefore, a directional antenna would help direct the transmitted RF to the intended receiver, and it would assist in concentrating the RF energy in that direction.

However, there are situations where there may be several intended receivers that are not in the same direction, so an omni-directional antenna may be what is used. Figure 15 shows a comparison between an omni-directional antenna and a directional antenna.

As RF energy travels through free space, these waves encounter natural objects in the environment and human-constructed objects made from various materials. When a radio wave faces these obstacles, transmissions will be affected—in some cases, positively and others adversely.

An example of a positive effect is a signal reflected off the ionosphere, and then the earth's surface provides a greater communication distance for HF transmissions. A negative impact is when a signal is prevented from reaching its destination, as with mobile telephony, when a user cannot send or receive UHF smartphone signals from a concrete and steel-constructed underground parking garage.

Frequency impacts the rate at which *data*—digital information—is transmitted. As a general proposition, the higher the frequency, the faster data can be exchanged.[67] At HF frequencies, data transfer is much slower than, for

67. Marko Suojanen, *Military Communications in the Future Battlefield*, p. 96.

instance, at the frequencies where a smartphone or satellite Internet operates. In the latter case, data can be sent at speeds that allow real-time video and audio to be received. This phenomenon helps explain why mobile/cell telephone providers have transitioned to higher frequencies.

Frequency also impacts the distance a signal can travel. Frequencies above 30 MHz are used for applications that can establish line-of-sight transmissions. Where line-of-sight (known as *ground waves*) is not possible, frequencies below 30 MHz are used. Having said that, 30 MHz is a "line in the sand" that is observed as the transition point.

Atmospheric conditions do affect transmissions on either side of this sand line. For example, distances of many thousands of kilometers have been achieved on frequencies of 50 Mhz, and to a less degree on 146 MHz. But as the frequency increases, say when we reach 450 MHz, these atmospheric anomalies disappear, and it is strictly line-of-sight.[68]

The same applies to the HF part of the spectrum as the frequencies drop. Around 3.5 MHz the distance for reliable communication decreases to between a few hundred to a few thousand kilometers. It would be fair to say that at about 14 MHz it is possible to have dependable world-wide communications.

The reason frequencies in the HF bands affect signals at different frequencies is because the ionosphere is comprised of multiple "layers." These layers vary in

---

68. Ian Poole, editor, *Antennas for VHF and Above* (London: Radio Society of Great Britain, 2013).

distance from the earth's surface and are identified by the way the sun charges the atmospheric particles.

Physicists have labelled these layers D, E, F1, and F2. Table 2 shows the approximate boundaries of these layers. These distances are not precise but are relative indicators of where different HF frequencies can be expected to be refracted.[69] This is because the ionization of the ionosphere changes from day to night, and in response to the sun's activity (i.e., the number of sunspots).[70] Generally speaking, the more sunspot activity, the better the band conditions.[71]

Table 2—Ionospheric layers.

Layer	Approximate Distance
F2	> 210 km
F1	160–210 km
E	80–160 km
D	60–80 km

The data in Table 2 indicates that if a radio signal can be refracted by the upper F1 and F2 layers, it will have the potential to travel a longer distance than if refracted by the

69. Known as the *maximum usage frequency* (MUF). This is the upper frequency where a radio wave will no longer be refracted. The MUF varies in response to the sun's activity, time of day, and season of the year.

70. Also, the seasons affect ionospheric propagation because of the Earth's change in axis. See Figure 17.

71. The number of sunspots follows an eleven-year cycle between successive peaks in activity.

D layer.[72] These waves are called *sky waves,* and the distance is called the *skip zone*.[73]

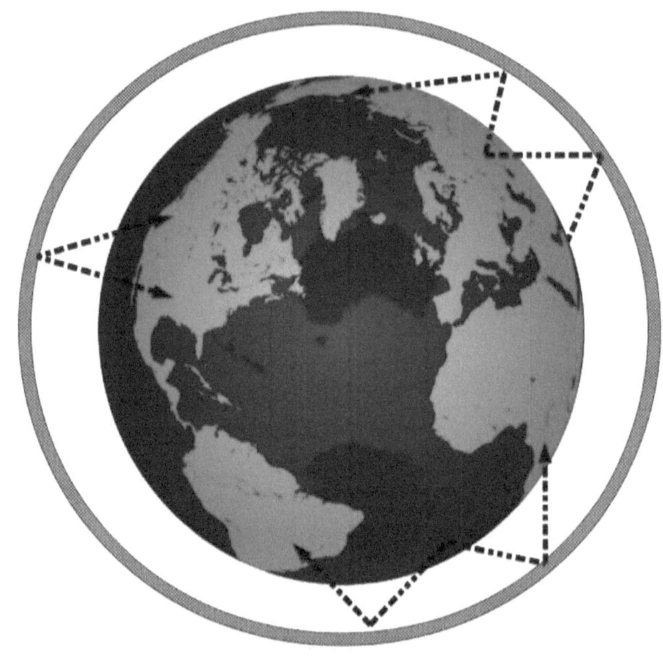

Figure 16—Diagram showing single (left) and multiple "hops" (right top and bottom). Courtesy of Wikipedia/kf4yfd.

Why does this phenomenon occur? Because the F2 layer allows the angle formed further from the earth's surface (i.e., *critical angle*) to reach a greater distance. At, say, 3.5 MHz (i.e., the 80m band, which is situated just above

72. Eric P. Nichols, *Propagation and Radio Science* (Newington, CT: ARRL, 2015), pp. 6-16 to 6-20.

73. See Figure 16 for an illustration of sky waves being reflected off the ionosphere. The area under the reflected signal is known as the *skip zone*.

the AM broadcast band), the D layer may allow reliable communications during daylight hours of around 400 km.

At night, when the D layer losses its ionization, and hence its ability to refract, 80m band signals are likely to travel one to two thousand kilometers using the F layers. This explains why during the day, only local and regional AM radios are heard, yet when night falls, interstate stations can be heard.

Another variable worth mentioning is the antenna. A resonant antenna at a half wavelength above the ground will help maximize an operator's chances of getting the best performance under these varying conditions. Nonetheless, a smaller, compromise antenna at a lower height will perform but not as well, thus jeopardizing reliability.[74]

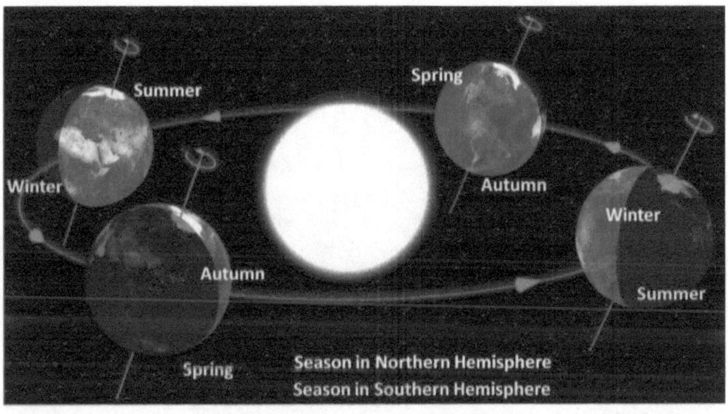

Figure 17 In addition to the sun's solar activity, HF propagation is affected by the seasons. Courtesy of Wikipedia.

---

74. A risk assessment needs to be considered if the communications link is "mission critical." See, Hank Prunckun, *Methods of Inquiry for Intelligence Analysis, Third Edition* (Lanham, MD: Rowman & Littlefield, 2019), pp. 201–208.

To recap, the sun's activity influences how the ionosphere behaves at different times of the day and night, as well as at each of the four seasons (i.e., caused by the inclination of the earth's axis as the earth orbits the sun). The D through the F2 layers react to the sun's activity, so the MUF can be estimated to ensure that a radio operator can get their message through. The reason for this—other than the technical issues of battery power and longevity, and RF power—is propagation.[75]

---

75. Eric P. Nichols, *Propagation and Radio Science* (Newington, CT: American Radio Relay League, 2015).

— CHAPTER SIX —

ELECTRICAL SAFETY

As a general proposition, low-voltage DC circuits don't pose a safety issue. That is to say, experts advise that voltages below 48 volts DC with a current of 15mA or less are usually safe.[76] Once these specifications are exceded, the electrical safety literature says DC voltages become dangerous. However, all mains voltages, which can be 120VAC or 240VAC in single-phase wiring, depending on the country, are lethal.

Each country has a national wiring code that identifies each wire in a cable by color. These colors are green or green-yellow (or bare wire) for the earth, red or brown for active, and blue or white for neutral. These colors cover most applications, but there are variations from old codes to new codes, as well as country variations (due to imported electrical devices), so you must check your local electrical codes to be sure.[77]

The reason for color identifying the different phases of a mains cable may be obvious to some, but let's state it explicitly—to avoid shock, burn, explosion, and fire. If

---

76. Having said that, a low 12-volt but high current circuit—as in the case of a car battery—can cause thermal burns if, for example, you short the terminals with a ring on your finger.

77. David Marne and John Palmer, *McGraw Hill's National Electrical Safety Code (NESC) 2023 Handbook* (New York: McGraw Hill, 2022).

the color pattern of is not observed and a device is wried incorretly, it could result in electrocution.

The next aspect of electrical safety we need to examine is bundling the three wires into a cable. We'll start from the wall socket where radio equipment will be connected. The first precaution is that the socket in which the plug is inserted needs to be cracked or broken free. If faulty, it may fall apart in your hand as it is inserted. Also, ensure that three conductors on the plug are securely fixed and that the cable has a strain relief grip firmly holding the cable in place. This will make sure the wires are not dislodged when you go to pull the plug out of the wall socket.

The important issue is the mains power cable is that it must be robust to withstand wear and tear associated with radio use, and it must be waterproof. It is not a good idea to run cables across walkways; apart from creating a trip hazard, the cable becomes susceptible to damage from foot traffic. Cables in walkways need to be covered with purpose-made cable protectors.

Sometimes it is necessary to feed several pieces of equipment from one wall socket. This can be done by using a four-way extension block, provided the block feeds only low-power equipment. Check the specifications when you buy an extension block. If your current demands are greater than this, feed it directly from the socket. "Daisy-chaining" extension leads is not a good idea; invest in a cable that is adequate for the distance.[78]

As for the cable itself, arguably the most important wire of a mains cable is the green–yellow earth wire. This wire

---

78. Rob Zachariason, *Electrical Safety* (Boston: Cengage Learning, 2022).

is the backbone of electrical protection systems. It acts to direct any short circuit to earth, thus avoiding shock, injury, or death.

In the case of radio equipment, check to ensure that the earth wire is connected to the radio's chassis or case directly after entering the enclosure. The wire should be fed through the eyelet of a lug and firmly fitted in position. The lug should, in turn, be bolted to the chassis using a shake-proof washer. What this does is form a "shield," protecting you from any of the mains voltage inside. If any metal components extend through the case (e.g., potentiometers), but are not in contact with the case, they need to be earthed also.

Otherwise, if you have a situation where there is a short, anyone touching that component could receive an electrical shock. In addition, the power cable entry point deserves some attention to prevent abrasion of the insulating covering of the wires that can result in contact with the chassis. The common precaution comprises a thick rubber grommet through which the cable passes. Once inside the case, this cable is held securely with a plastic cable clamp.[79]

Electric shock is produced by a current passing through the body. The electric current has a particularly devastating effect on the body's nervous system, but can cause burns to skin, eyes, and other soft tissue areas, either from the direct passage of current through the body or from contact with a surface which has been heated by the current. Minor shocks can cause injury after involuntary muscle contraction. Fires can also be caused from

---

79. Phil Simmons, *Electrical Grounding and Bonding, 6th Edition* (Boston: Cengage Learning, 2020).

electrical sparks, short circuits, or heating which are the result of overloaded circuits.

The result of electric shock is dependent on the strength of the current involved—this is based on the voltage and the body's resistance. Voltage is a straight-forward measurement; in North America, it is 120V, and in Europe, Australia, and New Zealand it is 240V. The body's resistance is the path the electricity flows.

If you were to touch a live wire with your hand, the current would flow from your hand, up your arm, down your torso, continuing down your leg, and finally to earth from your foot. Or, if you were using both hands, it would flow from, say, your right hand, up your arm, across your chest—and through your heart—down your left arm and hand to earth. Although the second example is a shorter path with less resistance, it is more dangerous than the first. Medicos say that death can result from contact with a normal 240V house supply when the amperage is greater than 30mA, and this only has to be for a period as short as 40ms.[80]

---

80. R.C. Lee, E.G. Cravalho, and J.F. Burke (editors), *Electrical Trauma: The Pathology, Manifestations and Clinical Management* (Cambridge: Cambridge University Press, 1992).

— CHAPTER SIX —

OPERATING PROCEDURES AND PRACTICES

We have covered the essential science and mathematics that underpin spy radio, but there is another side to understanding the tradecraft, and that is operating the equipment and on-air practices. Arguably these methods are as important as the technical side of any operation. Many on-air procedural errors have given opposition forces insight into the secret message traffic that have proved as devastating as if the operative handed over the technical know-how of the equipment they used.

We have covered a few spy radio applications, from clandestine broadcasts to microwave point-to-point transmissions. With each use comes a set of operating procedures and practices. Let's first look at the deceptive methods of the large extra-legal broadcasters.[81]

Clandestine broadcasting shares many of the same on-air methods as its legal cousin. For all intents, when a listener hears one of these stations, it will sound like a normal AM or FM broadcast. There will be an announcer who may even provide their name (whether it is their legitimate name needs to be questioned), there will be a regularly announced station identification, and with this,

81. The term *extra-legal* refers to stations that "...operate in violation of international laws or treaties, or in violation of national laws." Harry L. Helms, *How to Tune the Secret Shortwave Spectrum*, p. 7.

there could be the station's location (again, these details are likely to be deceptive, as well as the station's sponsor).

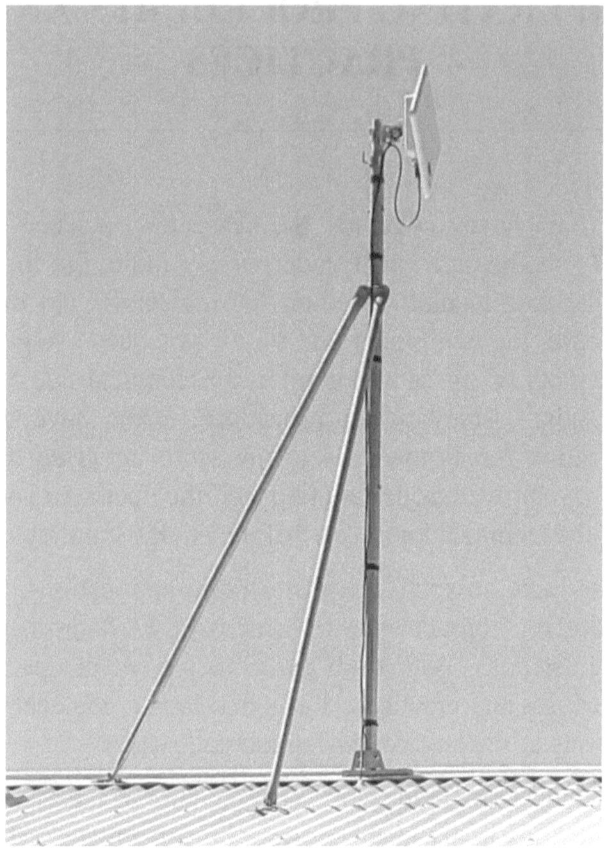

Figure 18—A microwave point-to-point antenna. Author's collection.

If it is a music station, there will be a playlist of artists and bands. Yet, above all, because clandestine broadcasts' messages are to incite political change or civil action in the targeted country, their transmissions will not be hidden—they will be promoted, announced, and widely circulated. Still, unlike a legitimate broadcaster—take, for example, the British BBC—the station will have a finite life. The station's lifespan will correspond with the political situation that is the subject of the messaging, so clandestine stations "...are born and die according to changing political fortunes."[82]

Although clandestine broadcast stations hide their true supporters, it is not possible to completely hide from being physically located. Even radio hobbyists can use direction-finding techniques to pinpoint the station's antenna (recall the studio could be anywhere in the world if the transmitter site has an Internet connection). A military or government intelligence agency, could, without doubt, locate the transmitter as well as determine many other aspects of the station's technical operation.

Regarding this last point—military and government intelligence capabilities—it is important to note that because of their high level of technical expertise and the state-of-the-art equipment they use, other forms of spy radio need special operating methods to perform as intended—that is, secretly.

Let's examine land, sea, and aeronautical mobile radio communications. Often, these mobile stations communicate with a base station (e.g., a command or control center), so we will include these fixed station operations in our discussion.

What we are referring to as a mobile station is any handheld or vehicle-mounted transceiver. A fixed station is where the transceiver and its antenna are not transportable. These descriptions are a guide, there can be situations where, say, a vehicle-mounted station is parked, and a temporary antenna erected, or backpack carried

---

82. Harry L. Helms, *How to Tune the Secret Shortwave Spectrum*, p. 8.

equipment so that the station becomes a *portable* station.[83] Other variations are possible, but we will work with these descriptors.

Figure 19—An example of a portable station. Courtesy of the American Radio Relay League.

If many stations operate on a particular frequency, there must be a protocol so that one station knows who the other stations are. In landline and mobile telephony, we use telephone numbers. In radiotelephony, we use callsigns.

In the commercial realm, there is a government department that oversees telecommunications license stations and issues callsigns. Military and law enforcement agencies also use callsigns. A police dispatcher will issue a request for a patrol to respond to a call for public assistance by using the callsign of the

---

83. See, as an example, Stuart Thomas, *Portable Operating for Amateur Radio* (Newington, CT: ARRL, 2018).

patrol. A military aircraft landing will communicate with the tower using its callsign. Callsigns are used by commercial companies, for instance, public bus operators, when drivers contact their base of operations.

Callsigns can comprise letters, numbers, or a combination of both, or they can be names or phrases. When using letters, the NATO phonetic alphabet is used. The reason is that it avoids mistaking one letter for another, especially if the operator has a regional or foreign accent. This is the NATO phonetic alphabet:

> **A**lfa, **B**ravo, **C**harlie, **D**elta, **E**cho, **F**oxtrot, **G**olf, **H**otel, **I**ndia, **J**uliett, **K**ilo, **L**ima, **M**ike, **N**ovember, **O**scar, **P**apa, **Q**uebec, **R**omeo, **S**ierra, **T**ango, **U**niform, **V**ictor, **W**hiskey, **X**-ray, **Y**ankee, **Z**ulu.

In intelligence work, a station's callsign may be disguised to make it difficult for an opposition force to create a diagram of what is called a *radio net*. A radio net is any group of radio stations operating in a complementary way. It could be a flotilla of ships, a police district's fleet of mobile patrol vehicles, or on a larger scale, an army division's radio network comprising, at the base, unit-to-unit communications (i.e., sub-nets with *tactical call signs*[84]), moving up each echelon to headquarters level.

Undercover operatives may have randomly selected callsigns so that the call sign if intercepted, is not an obvious giveaway that describes the mission underway.

---

84. Radio traffic is a source of intelligence. One particular type of analysis is that of callsigns. Therefore, to make it difficult of an opposing force, military units use tactical callsigns that are changed daily. Law enforcement and other paramilitary agencies also use tactical callsigns. U.S. Army, *Infantry Division Operations: Tactics, Techniques, and Procedures, FM 71-100-2* (Washington, DC: Department of Defense, 2020).

The common practice, contrary to what we see in cinemas, is for the station calling another station is to announce the desired station's callsign first, followed by their callsign.

Starfish, this is Prime, over.

Once heard, the receiving station would reply:

Starfish, send.
Then, Prime would send its message.

The reason for this sequence is that the calling station what's to attract the desired station's attention. Sometimes, the desired station's callsign is announced several times, followed by the sending station's callsign, which could also be repeated several times. This is common where the call is being made "cold," that is, when the transmission isn't anticipated (i.e., when stations are on standby).

If dialogue has been established and the two stations' conversation is underway, the stations may dispense with the use of callsigns. Or, the stations may abbreviate the calling procedure:

Starfish, Prime.
Starfish.
Followed by Prime's message.

At the end of the exchange, the calling station concludes with the word, *out*.[85] Radio operations has its own

85. Note that it the phrase *over-and-out* is not used because *over* means the station is handing the microphone back to the other station, and *out* means the end of the conversation. It is either *over* or *out*.

lexicon of procedural words—referred to as *prowords*. The more commonly used words are listed below.

Copy—The station has received another station's transmission. Note it is not the same as *Roger*.

Do you read—Asking the other station(s) whether they can hear the transmitting station.

I read back—The receiver's response to the sender's request to "Read Back."

Out—Indicates the end of the transmission. There is no response from the other station(s).

Over—A station's transmission has finished, and it expects the other station to respond. Sometimes the proword, *back*, shorthand for *back to you*, is used in place of *over*.

Radio check—The station is asking what its signal strength and readability are.

Read back—Repeat the message I just transmitted.

Roger—Means affirmative. Sometimes *Romeo* is used in its place.

Say again—Please repeat your message.

Send—Begin your message.

Standby—Pause for a few seconds.

This is…—Announcement of the transmitting station followed by its call sign.

Wait—Same as *standby*.

Wait out—Pause for longer than a few seconds.

Wilco—I *will comply*.

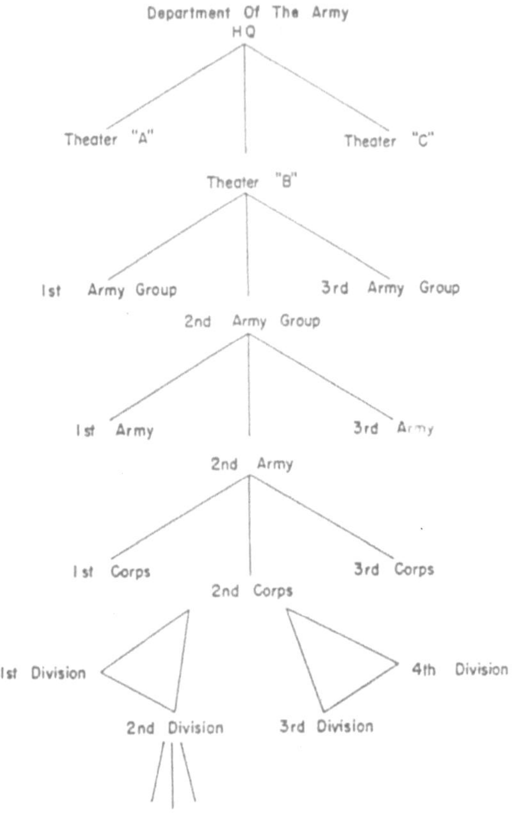

Figure 20—Diagram of a typical army radio net. Courtesy of the U.S. Army.

When it comes to one-way transmissions where the receiver does not acknowledge or cannot acknowledge reception, the procedure will rely on a schedule. It is arranged that the sender will transmit its message at a certain time on a given day, with the receiver monitoring the frequency. This is referred to as *blind sending*.[86]

---

86. U.S. Department of the Army, *Fundamentals of Traffic-Analysis (Radio-Telegraph), TM-32-250*, first published in 1948

Variations are, of course, used. For field operatives to transmit their messages, there may be several times each day or several days each week with several times to cater for unforeseen in-country circumstances. Variation may also be required when, for instance, a satellite will pass overhead for the operative to establish an uplink.

For a command-level base-station to communication with the field, the same arrangements can apply—days and times established. If the messages are sent blind, the message may be transmitted at several times in case the field operative was not able to receive the first of the messages.[87] This has been observed with so-called numbers stations.

Let's take a hypothetical situation where the different operational elements we've been discussing come into play.

A case officer sets up a clandestine agency meeting at a café located in an overseas country. The case officer enlisted the help of a countersurveillance team to ensure that the opposition wasn't observing the meeting. The case officer used a micro-miniature microwave frequency transceiver worn in her ear to keep in contact with the surveillance team. The device is like a hearing aid in that it has no wires protruding from the small apparatus. (Figure 21 is a portrayal of what this casual contact might look like).

---

and reprinted by Aegean Park Press, Laguna Hills CA, no date, p.103.

87. This could be because of in-country issues but also because of atmospheric noise and/or adjacent frequency interference from other band users.

Figure 21—A depiction of a field agent meeting with her case officer. Courtesy of Microsoft.

From what we already learned; we know that the device would have a tiny battery capable of perhaps only a few hours of operating time. Time will be reduced if the device operates in full-duplex mode, transmitting and receiving simultaneously, like when we use a telephone. The distance it will be able to transmit will also be very short—maybe a few meters.

If there are people at a forward command post (e.g., a van parked nearby) to record the meeting or intervene by issuing instructions to the case officer, then the little ear-worn device will not do the job. Yet, knowing this, the case officer can opt to have the earpiece Bluetooth® a signal[88] to a handheld two-way radio in her carry bag that relays the signals to and from the surveillance operatives and the forward command van on a VHF frequency using higher power.

Let's take this example one step further. The signals from the café can then be relayed by the forward command

---

88. Bluetooth® operates on the frequency range of 2.402 GHz to 2.48 GHz, which is in the upper-UHF part of the spectrum.

van to an international command center via, say, a microwave link to the Internet (e.g., by using a smartphone), or via a satellite telephone or transceiver. But the tiny earpiece can't do more than that, despite what is portrayed in the cinema.

Lastly, we'll discuss distress calls. A distress call consists of the distress proword applicable to the mode of transmission (voice or telegraphy), usually sent three times, followed by the words *this is* (or *de* in Morse code), and then the callsign of the station in distress, again up to three times.[89]

The distress proword for voice communications (i.e., radiotelephony) facing grave or immediate danger is *mayday* and for Morse code (i.e., radiotelegraphy), it is the three letters, *SOS* (e.g., … --- …).[90] Once contact is made with another station, the distressed station will advise what the danger it has encountered, its position, and the type of assistance needed.

Is this the limit of radio operating procedures? No, but this brief explanation of how on-air practices are performed, provides sufficient understanding of on-air mechanics of spy radio.

---

89. Australian Maritime College, *Marine Radio Operators Handbook* (Launceston, Tasmania: Australian Maritime College, 2003), pp. 29–36.

90. See Appendix B for a table of letters and numerals displayed in International Morse code.

— CHAPTER SEVEN —

COUNTERMEASURES

As cleaver as spy radio engineers and technicians are in devising defences against discovery and interception, an opposition force's technologists will come up with methods or devices to counteract them. These approaches are known as *countermeasures*.[91] These are divided into two types—defensive and offensive, and have their theoretical base in the abbreviation—CIA.

These three letters are *not* short for the Central Intelligence Agency but for the confidentiality, integrity, and accessibility of wireless systems and the traffic they carry. Confidentiality is concerned with secrecy, whereas integrity is about reliability, and availability is about access.[92]

Defensive measures are intended to protect one's own radio equipment and signals, while offensive countermeasures are meant to interfere, block, or otherwise disrupt the opposition's transmissions and ability to receive signals. In either application, countermeasures can be passive and/or active.

---

91. Hank Prunckun, *Counterintelligence Theory and Practice, Second Edition* (Lanham, MD: Rowman & Littlefield, 2019).

92. Adapted from, Bruce Schneier, *Click Here to Kill Everyone: Security and Survival in a Hyper-Connected World* (New York: W.W. Norton, 2018), p. 79.

Figure 22—Monitoring the electromagnetic spectrum. Courtesy of PeakPx.

In Chapter One we discussed one-way voice links (OWVL) in relation to numbers stations as a method of guarding a field operative's identity and location. This method can be considered a passive defensive technique because it is a security function—it protects the person receiving the signals by not requiring them to transmit, thus eliminating a signal that can be traced to the sender.[93] Thus, it fulfills the C in the CIA theory—it helps maintain confidentiality.

The other protective aspects of OWVL are that they are transmitted at a time and frequency that is difficult to detect. That is, unless an opposing force is simultaneously monitoring a wide range of frequencies across many bands, the OWVL might be missed.

---

93. Possession of a shortwave radio also offers the agent "plausible denial" should it be found in their passion.

If, however, the opposing force is using a broad-spectrum radio receiver, such as a software-defined radio (SDR), which is programmed to automatically record a numbers station once it is detected. Automatic detection and recording can be done via an artificial intelligence software application that senses the hallmarks of a numbers station.

If it is likely that an opposing force has the technical and financial wherewithal to accomplish interception in such a manner, the "message" can be digitally encrypted, adding another layer of confidentially.

Yet, knowing that an opposing force will be employing some form of interception techniques, active countermeasures can be added to disrupt its collection efforts. How might this happen? One method that immediately presents itself is to flood the airwaves with fake signals to non-existent field operatives.[94] In this way, the intercepting station will be using recording equipment and employing their finite (and expensive) computer resources to try to decipher the messages. In effect, these fake transmission act as decoys, obscuring the real signals.

Related to encryption is the passive defensive countermeasure of *scrambling*. This technique uses an algorithm to distort the modulation on the carrier signal, so the audio is unrecognizable to anyone monitoring. A scrambler inverts the lower speech frequencies, so there are higher, and the higher frequencies are lower.[95]

---

94. Known as *simulation* (i.e., "ghost operatives") and via *manipulation* (e.g., false traffic levels). Jon Latimer, *Deception in War* (Woodstock, NY: The Overland Press, 2001), p. 78.

95. John M. Carroll, *Secrets of Electronic Espionage* (New York: E.P. Dutton, 1966), pp. 203–206.

Figure 23—Al-Qaeda did not stand a chance against wireless interception during the so-called Global War on Terror.

As with every countermeasure, there is a potential solution. In this case, the intercepting station can use a de-scrambling device to solve the sound splitting to unscramble the audio. This was done by the Nazis during the Second World War. From a listening post in occupied Holland, the Germans intercepted and de-scrambled the radio-telephone conversations between Prime Minister Churchill and President Roosevelt.[96]

There is an axiom that states the longer a station is on the air, the greater the likelihood its signals will be intercepted. A passive defensive technique to help maintain confidentially is to use *burst transmissions*.

---

96. Bruce Norman, *Secret Warfare: The Battle of Codes and Ciphers* (Melbourne: Wren Publishing, 1973), pp. 154–155.

The method uses a system of recording the message and then compressing the time so that, for example, a few minute's worth of message is condensed into a few seconds or less. The method is analogous to playing a commercial music recording at, say, 128-times or faster than the recorded speed. The receiving station then plays the recording at the required, slower, speed.

Although not a perfect defense, it limits the time an operative needs to be on the air.[97] Although this might appear to be an impossible task to intercept, some nations' intelligence agencies can do this. When it comes to lesser sophisticated organizations, they have little defence against interception. By way of example, al-Qaeda advised its operatives to limit the time of transmissions to less than "…five minutes in order to prevent the enemy from pinpointing the device location."[98]

Five minutes?! I would suspect that it would take less than a few sections for a listening post to identify and locate the signal (direction-finding) as being of interest and automatically start a recording of the exchange. It would probably take another few seconds to have the artificial intelligence-based system flash an alert across the computer monitor of an intelligence analyst, who would then initiate a series of offensive actions that were stipulated in their agency's protocol manual or its standing orders.

With such a pathetic countermeasure against opponents using state-of-the-art equipment, it is no wonder that al-

---

97. Burst transmissions can be combined with scrambling or encryption.

98. *Military Studies in the Jihad Against the Tyrants: The Al-Qaeda Training Manual*, p. 39.

Qaeda did not stand a chance of success during the so-called Global War on Terror.

Figure 24—A tactical force forward observer in Afghanistan (July 2010) using a transceiver that employs a point-to-point satellite antenna. Courtesy of the U.S. Department of Defense. Photograph by Pfc. Chris McKenna, U.S. Army.

Another form of burst transmission uses devices designed for short-range agent communications (abbreviated as SRAC).[99] These devices allow agents and their case officers to exchange information without the need to get too close to each other, or resort to tradecraft that requires

---

99. Wallace and Melton with Schlesinger, *Spycraft*, p. 478.

them load-and-clear a "dead-drop"[100] or conduct a "brush-past."

SRAC devices removed the risk that the passing of physical information will be observed because of their ability to transmit and move. Though, it does not eliminate that the opposition will anticipate this manoeuvre and have an intercept device ready to capture the communication.

An interesting variation of an agent–handler SRAC exchange is where an agent pauses at a designated spot and speaks aloud (as if they are talking to themselves, but more subtly) their message, then moves on. In the listening vicinity will be a hidden transmitter that relays the agent's message to a covert listening post nearby. This "listening post" could be as elaborate as a dedicated room in a nearby embassy or as simple as a receiver and digitally recorded in a person's shopping bag.

On a more sophisticated scale, the SRAC are point-to-point transmissions (see Figure 24). These can be effective using frequencies in the upper end of the UHF spectrum and into the SHF range (e.g., microwaves). As we know, these signals can be narrowed to form a tight angle that leaves little or no radio signals outside its precise focus.

To help visualize this, we'll look at the analogy of a satellite television dish. We see that if a roof-mounted dish is moved in any direction other than where it is aimed directly at the satellite, the receiving signal strength drops.

In this case, the satellite transmission is aimed to cover a large footprint on the earth, perhaps as large as focused as

---

100. Known as a *dead letter box* in the parlance of the British secret services and *taynik* to former-Soviets spies.

two or three European countries or expanded to cover an entire continent. However, if the satellite's antenna were designed to concentrate its signals in the way the house-mounted TV dish does,[101] we now have a point-to-point link which's signal will be practically impossible to be intercepted unless the eavesdropper is in the line of transmission (or has their own satellite . . .).

Burst transmissions are akin to the tactic of "transmitting and moving." The theory behind this countermeasure is, as stated in the al-Qaeda manual, to limit the time on the air so that direction-finding techniques cannot be employed. But this passive defensive measure will only work if the opposing force is restricted in their level of technology. The United States, European NATO countries, and the so-called Five Eyes countries[102] have access to signals intelligence from aircraft- and space-based receivers that can locate a transmitter within a short time (as do the Russian and Chinese signals intelligence agencies[103]).

Spread spectrum techniques are variants of the burst method. First used in the 1940s, this procedure requires the sender's equipment to send part of the message over one frequency, another part on another frequency, and so on throughout the transmission. "Hopping" between

---

101. Known as a *Yagi-Uda* antenna, or *Yagi* for short. The name comes from the design's inventors (1926), Shintaro Uda of the Tohoku Imperial University, Japan, and Hidetsugu Yagi.

102. i.e., Australia, Canada, New Zealand, the United Kingdom, and the United States.

103. Matthew Aid, "All Glory and Fleeting: SIGINT and the Fight Against Terrorism," in Christopher Andrew, Richard J. Aldrich, and Wesley K. Wark, editors, *Secret Intelligence: A Reader* (London: Routledge, 2009), p. 68, and Nigel West, *Historical Dictionary of Signals Intelligence*, pp. 57–58.

frequencies lowers the probability for interception,[104] as does turning the carrier on-and-off (i.e., pulsating) in a "time hopping" fashion which is based on a pseudorandom generator.

Figure 25—U.S. Department of Homeland Security factsheet.

Now let's look at a few offensive countermeasures, starting with *jamming*. This technique is intended to obstruct reception. During the Cold War, jamming was

---

104. Marko Suojanen, *Military Communications in the Future Battlefield*, pp. 73–74.

used strategically by the Soviet Union to block foreign Western broadcast stations.[105]

As a tactical method, jamming can be used to, for instance, prevent cell phone signals in a specific location, as in the case where a terrorist incident is taking place. It can be applied to cases of espionage where an opposition agent is using the practice we discussed before of taking aloud so that a hidden transmitter to relay their message.

When it comes to protecting information that contains references to intelligence methods and sources, a passive system known as a *SCIF* is used. SCIF stands for sensitive compartmented information facility, which can be a portable apparatus (e.g., for use in a hotel room) or a permanent structure (e.g., the White House Situation Room). The facility offers extremely high safeguards against visual, acoustic, and electronic eavesdropping. There are standards for constructing SCIFs, such as the U.S. National Counterintelligence and Security Center's *Technical Specifications for Construction and Management of Sensitive Compartmented Information Facilities*.[106]

Leaving aside the visual and auditory aspects of a SCIF, we will examine the anti-electronic characteristics, which are based on a "Faraday cage." In 1836, the British

---

105. During the Cold War, the Western view was that jamming was a violation of the principle of freedom of information—a fundamental human right. Rochelle B. Price, "Jamming and the Law of International Communications," in Michigan Journal of International Law, Volume 5, Number 1, 1984, p. 391–393.

106. National Counterintelligence and Security Center's *Technical Specifications for Construction and Management of Sensitive Compartmented Information Facilities, Version 1.5* (Washington, DC: , Office of the Director of National Intelligence, 2020).

physicist Michael Faraday designed and build a container that protected devices placed inside from electric fields.[107]

It works by covering the six sides of the space under protection with conductive material. The theory is that any externally applied electrical field will result in the charge being distributed throughout the cage's shielding that in turn, cancels the electrical field on items inside the cage. It works in reverse also to stop electric fields inside the cage from "leaking" out.

Figure 26—A Faraday bag protects credit cards and other items from electronic theft. Courtesy of Fullmetalwikial.

---

107. Nancy Forbes and Basil Mahon, *Faraday, Maxwell, and the Electromagnetic Field: How Two Men Revolutionized Physics* (Amherst, NY: Prometheus, 2014).

A SCIF is useful when the users are confident that there are no listening devices inside. Where there is a chance of a hostile device, a technical surveillance countermeasure (TSCM) "sweep" can be conducted. Because a "bug" emits electromagnetic radiation, a wideband receiver[108] is used to detect the radio waves. Sweeps are good but can be circumvented by switching the device off. Hence, a good sweep will combine an electronic search and a physical inspection of the environment.[109]

---

108. A wideband receiver is used because it is unknown what frequency or band the hidden transmitter will be operating on. A single band receiver is more than likely to miss the signal if the incorrect band is being listened to.

109. Scott R. French, *The Big Brother Game* (Secaucus, NJ: Lyle Stuart Inc, 1975).

— APPENDIX A—

TWELVE COMMON ELECTRONIC COMPONENTS

Batteries—store large electrical changes to power devices and circuits.

Capacitors—capacitors store small electrical changes as a means of producing a time-delay in a circuit.

Circuit Breakers—offers circuit protection by stopping the flow of current due to overload.

Diodes—a component that restricts the flow of current in one direction.

Fuses—perform the same function as a circuit breaker but need to be replaced, whereas a circuit breaker can be reset.

Inductors—a passive component that stores magnetic energy in a circuit. Also called a *coil* or a *choke*.

Integrated Circuits (IC)—often called "chips", are groups of components imprinted onto a layer of semiconductor material.

Relays—switching devices that use electromagnetism to open and close a circuit's input and output contact(s).

Resistors—control voltage and current in circuits.

Switches—a device that connects or interrupts a current flow in a circuit or diverts the flow from one circuit to another.

Transformers—a device that transfers electrical energy between circuits and, in doing so, changes the voltage, current, or phase.

Transistors—devices comprising semiconductor material that to provide either amplification, detection, or switching.

— APPENDIX B—

INTERNATIONAL MORSE CODE

When and radio operator sends Morse code, the formation of letters and numbers follows the convention below[110]:

- The length of a "dot" is one unit.
- A "dash" is three units.
- A space between of the same letter is one unit.
- The space between letters in three units.
- The space between words is seven units.

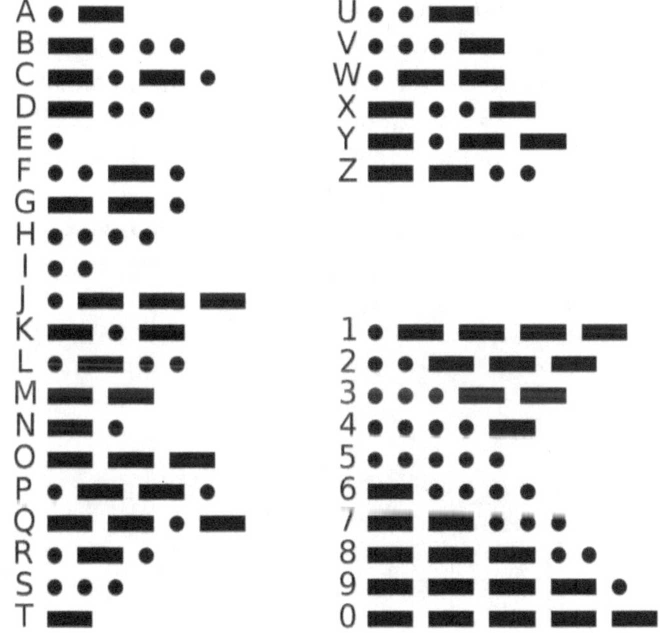

---

110. As advocated in, American Radio Relay League, *Morse Code Operating for Amateur Radio: 1* (Newington, CT: ARRL, 2013).

ABOUT THE AUTHOR

Dr Henry (Hank) Prunckun, BSc, MSocSc, MPhil, PhD, is an Adjunct Associate Research Professor with the Australian Graduate School of Policing and Security, Charles Sturt University, Sydney. He is a former Australian government intelligence analyst who spent much of his twenty-eight-year operational career in tactical intelligence and strategic research but also served in security, investigation, and counterterrorism. Dr Prunckun holds an Advanced amateur radio licence, which allows him to build and modify radio transmitting and receiving equipment. During his operational career, he was conferred with two literary awards and a professional service award by the International Association of Law Enforcement Intelligence Analysts. After retiring from government service, he initially worked as a freelance private investigator, then spent almost a decade-and-a-half as a research criminologist studying transnational crime—espionage, terrorism, drugs and arms trafficking, and cyber-crime.

INDEX

www.ingramcontent.com/pod-product-compliance
Lightning Source LLC
LaVergne TN
LVHW051017080826
845145LV00009B/2670

* 9 7 8 0 6 4 5 6 2 0 9 1 7 *